Everyday Science, Maths, and Engineering

First Edition

By Isaiah Watson

Prologue

STEM, or Science, Technology, Engineering, and Mathematics. My personal pursuit in study is in the STEM field, and, let me tell you; it can be complicated and difficult at the best of times. But here, I hope I can help you learn about stem in an easy to digest, simple, engaging format. That being said, just because you've read my book, that doesn't mean you're an expert in all things science or a human calculator; rather, this book aims to give you the tools to learn how to think from a STEM perspective. I hope you enjoy.

Oh, and a side note; this isn't written in any particular order; I tried to write things in order of what they build off of, but I did break that by accident (Specifically, I accidentally forgot to explain loops until chapter 14).

Chapter 1: One, Zero, and Other Numbers

Let's start with a simple refresher of numbers, and how numbers work, since I will be referencing mathematics a lot during this book. In short, the Hindu-Arabic numeral system (that is used by most of the world, including England, the United States, Canada, and many other countries). We use 10 characters to denote various values from zero to nine. These characters are used to denote specific values multiplied to a power of 10 (more on that later) They are as follows:

- 0: zero, cero, null

- 1: one, uno, ein

- 2: two, dos, zwei

- 3: three, tres, drei

- 4: four, cuatro, veir

- 5: five, cinco, fünf

- 6: six, seis, sechs

- 7: seven, siete, sieben

- 8: eight, ocho, acht

- 9: nine, nueve, neun

It's worth noting that this isn't the only way a number system has to work; for instance, the romans used the following numerals

- I: Unum

- V: Quinque

- X: Decem

- L: Quinquaginta

- C: Centem

- M: Mille

However, their numeral system was more focused on addition than our exponential system; so instead of writing a number like 198 as 100+90+8, they would write it as 100+100-10+5+1+1+1, or CXCVIII. Notice how the 100's place and the 1's place have multiple characters representing them? Yeah, that doesn't really happen in Hindu-Arabic numerals.

But that's all old news, most likely; Chances are if you've picked up this book, you know how to read Hindu-Arabic numerals, and likely have some familiarity with Roman numerals (though if you do not, I would highly recommend learning both Roman and Hindu-Arabic numerals). What we're interested in here is actually a very interesting number system, which we will be referencing in various other locations; binary. Binary, thankfully, is written using the Hindu-Arabic numerals, however, not all of them; only "1" and "0", or on and off. I'll go into more depth later as to how this works, but for now, all you need to know is that binary is simply representing the same thing Hindu-Arabic numerals do, but in terms of powers of 2, instead of powers of 10; so for instance, the Hindu-Arabic numeral 52, in binary, would be 110100_2. Oh, also, the subscript "$_2$" just clarifies that a number is in binary, and when we learn hexadecimal, we will do a similar thing with that, by denoting it with the subscript "$_{16}$".

Anyways, now that we know what numbers are, what properties do they have? Well, let's go ahead and create some basic rules.

- For any number x, x=x. We will call this the numeral identity property.
- For any two numbers x, y where x>y, y<x. The reverse of this is also true. We will call this the numeral comparative property.

It's worth noting that I completely made those names up, but I based them on the names of actual mathematical properties which we will explore later. First, we need to add the ability to do an operation. We will explore counting first, since counting is the backbone of all mathematics.

Now, count to 9; I've already given you all the tools you need to count to nine; you know the names of the numbers from 1-9 in three languages now if you've been reading closely (English Spanish, and German) and also what the Hindu-Arabic numerals for such numbers look like. You see, numbers are sequential, and when numbers come in a sequence, we can iterate through that sequence and see new numbers. By the way, from this point on, we will be referencing numbers of two digits in length regularly; so be prepared.

Now, do me a favour and count from 3 to 6; notice how you can start counting from a point other than 0 (or 1). This is an important step; now that we can count from numbers other than the very beginning of our sequence, we can count a certain number of numbers higher or lower from another number. For instance, if I told you to count 4 numbers up from 7, you would end with 11; This, my friend, is addition. Let's define some properties, which will help us better understand how to use these operations later.

- Any number added to 0 will result in the original number: This is called the identity property of addition. The reason this works is that adding 0 simply means you are not adding anything, and since x=x, "x=x+nothing" is still x=x

- You can do sequential addition and still get the same result. This is called the commutative property of addition. In other words, 3+6+9 = 3+9+6 = 6+9+3, and so on and so forth

- The associative property says that even if addition is grouped in parenthesis, it results in the same thing. Really, this works the same as the commutative property.

Now that we have a solid foundation in addition and subtraction (which is really just backwards addition), we are ready for something a little more complicated; but first, I'd like to examine how this looks in binary, because it's rather similar to how it looks in the Hindu-Arabic standard system.

We can simplify addition with an "add" value and a "carry" value for each individual digit. For instance, if we add 35 to 27, the one's place adds to 12, so we add the carry value of 10 to the addition of the 10's place and the one's place becomes two. We then add the carry and the two original terms for the 10's place and get 60; and 60+2=62. Binary works similarly; for instance, let's look at the following:

$$1010 + 0011$$

For this, we add each digit together, and if they result in a number that cannot be expressed in one digit, we carry it over to the next. So here, we get a carry in the 2's place into the 4's place; and our end result is 1101. I promise, in later chapters, this will make a lot more sense. But for now, this ends chapter 1. In chapter 2, we will continue to explore addition, and also add in multiplication, with some practical usage as well.

Chapter 2: Addition, Multiplication, and Money

Okay, I concede that the name of the chapter is a tiny bit misleading; but we will talk about how understanding multiplication is integral to ensuring you're getting paid as much as you're supposed to. To start, however, we should refresh ourselves on how we got from counting to addition:

Addition is simply repeatedly counting up by one; similarly multiplication is simply adding up numbers repeatedly. For instance, it'd be a lot of work to write out 3+3+3+3+3+3+3+3+3 whenever you wanted to add three to itself nine times, so instead we can simplify it to be 3*9, sometimes expressed as 3 × 9 instead. Repetitive addition like this is where the concept of rates comes into play. By the way, the reverse of this is called division.

But before we can get into our rates, we need to first understand multiplication's properties; It has all the same properties as addition, except for a couple small changes; mainly, the identity property works off of 1 now, and anything times 0 is 0. To understand why, imagine multiplication as a chain of operations; if you only have one link, you have your link, and if you have no links, you have nothing.

Alright, time for money; if you've ever worked a job for someone else, you've probably received money for doing that task per hour. The "per hour" part means that every hour you work, you will receive that amount. So, let's look at an example.

Minimum wage in my city is 7.25/hr, and let's say I hire you to create ceramic pots for minimum wage (not that I'd ever do that, but still). Say you work 5 hours a day, for five days a week. That makes out equation look like this:

```
5 * 5 * 7.25
```

This evaluates to 181.25. Of course, some goes to tax, let's say 6%; so you receive 181.25*(100-6)/100. Now would be a good time to explain PEMDAS, or Pandas Eating Mayonnaise Don't Always Survive.

- P: Parenthesis are first. When doing an equation ,you do the inside of parenthesis before anything else

- E: Exponents are after Parenthesis. These we haven't gotten to yet.

- MD: Multiplication/division. These come at the same time, after exponents

- AS: Addition/subtraction. These come after multiplication and division.

So, now we have 181.25*94/100, which will tell us how much you end up making in the end (170.38). So far, pretty basic; but now let's add yet another dimension to things.

Multiplication allows us to examine our first interesting graphs; straight, diagonal lines. We will go over them in more detail in a later chapter, however, so hold off until then. Remember, so far everything has been based off of counting, that's because counting is the way we do operations. In the next chapter, however, we'll take a look at one more operation before we apply STEM to more complex topics.

Chapter 3: Exponents, Algebra, and Interest

I promise, this is the last of the chapters where I'm going to baby you, but for now we need to ensure that you understand this basic mathematical operation: Exponents. Exponents are simply repeated multiplication; just like how multiplication is repeated addition.

This allows us to compound multiple multiplications all at once, but in order to do that for any really useful purpose, we must first understand algebra. But before we get into algebra, let's define some properties of exponents:

- Multiplicative property: $x^y * x^z = x^{y+z}$. By the way, when you have two variables right next to each other, it means multiplication. Same when you have a constant next to a variable. I mention this here since later I will almost certainly use something like x^2x^5 in arithmetic explanations.

- Divisive Property: $\frac{x^{y}}{x^{z}} = x^{y-z}$. This is actually the same thing as the multiplicative property of exponents, since $x^{-y} = \frac{1}{x^{y}}$.

- Compounding exponents: $(x^{y})^{z} = x^{yz}$. This is, once again, derived from the multiplicative property of exponents, since

$$(x^{y})^{z} = (x^{y}) * (x^{y})... \text{ "z" number of times.}$$

- Identity property of exponents: $x^{1} = x$. This is due to the same reason as the multiplication property of identity; there is only one term here.

- Zero property of exponents: Anything to the power of 0 is 1. This isn't a super complex mathematical thing, it just means that if you don't multiply 1 by anything, you get 1.

Now that we know how exponents work, and some of their properties, we need to introduce one more element to make them worth the usage; Algebra. Simply put, algebra is mathematics that uses variables; or letters and symbols; in place of numbers. This is a hugely important step, because it allows us to introduce the dynamic of a parameter, a crucial step in designing models and algorithms. But I'm getting a bit ahead of myself, allow me to show you some equations:

$$x + 3 = 5$$

You can probably easily figure out that in this equation, x=2; but what may be a bit more complicated is an equation that looks like the following:

$$x = 2(x)(x + 1)^2$$

For this we need to understand how to solve equations like this. In order to do that, we want to isolate the variable, which we do by applying operations to both sides; the same operations to both sides to keep things balanced. Let's have a look at what that looks like in this instance. First, we would divide both sides by x, to give ourselves the

following:

$$1 = 2(x + 1)^2$$

Then, we would divide both sides by 2, and we would get the

following:

$$\frac{1}{2} = (x + 1)^2$$

From here, we square-root both sides (by the way, a root is simply

a backwards exponent; and the square root of a number x times itself

equals x). This gives us the following equation:

$$\sqrt{\frac{1}{2}} = x + 1$$

Finally, we subtract one from both sides and simplify, and we get:

$$\sqrt{\frac{1}{2}} - 1 = x \Rightarrow \frac{1}{\sqrt{2}} - 1 = x \Rightarrow \frac{1-\sqrt{2}}{2} = x$$

And now we can learn our first algorithm; an algorithm is simply a set of instructions which, when executed, perform a task. This task can be anything from recommending this book to you to reinforcing my addiction to purchasing new sets of polyhedral dice by having online stores shove them in my face. As someone studying IST, I have designed algorithms for computers before to perform basic tasks, but we will talk more about that in a later chapter. Firstly, we want to explore how to use algebra to solve an algorithm. Take the following, for instance; the compounded interest equation, the equation that determines how much the bank owes you in interest over a given period(or how much you owe them from your loans):

$$A = P(1 + \frac{r}{n})^{nt}$$

Let's quickly define the variables:

- A: This is the end amount, after interest is applied

- P: The initial amount, before interest is applied

- r: The rate of interest; the percent change each time that your
interest represents; this will be in terms of percent, so 1% would be
0.01, and so on and so fourth

- n: The number of times the interest is compounded per year (or any
other unit of time)

- t: The number of years (or relevant periods of time) from the
beginning to end

If we know four of these values, it becomes very easy to
determine what the fifth value is; and furthermore, if we know three,
then we can define the remaining in terms of each other. Let's have a
look at an example where we know four.

You happened to have a bank account with the Royal Bank of
Stemlandia; who will offer you an interest rate of 1% per season (3

months, making it happen 4 times a year). You have a vacation coming up to Stemmerland, and by the time you leave for vacation, you will have £10,000 in your account. You plan to be on vacation for two years and will pay for everything in cash, so that you don't need to spend any of your bank account. How much will you have in your bank account after those two years? Let's have a look at what this looks like in our equation:

$$A = 10000(1 + \frac{0.01}{4})^{2*4}$$

Evaluating the equation gives us $A = 10,201.76$, but you may have noticed something: The amount that it changed from one year to another increased; meaning that with exponents, we now have the ability to express acceleration. Recall how multiplication gave us the ability to talk in terms of flat rates; a one-dimensional, linear progression. Exponents allow us to explore increases at a second-dimensional level, meaning that now we can have rates increase.

Chapter 4: Modeling and Algorithms

We've seen examples of maths used to model simple scenarios, but you may be wondering how to use maths to model solutions pertinent to you; well, it all comes down to simplifying things and understanding the problem at hand. Let's look at an example.

I have a cat, she's a bit ravenous. She loves breaking her toys under the assumption that they're edible, even though she has a full bowl of food at all times. Let's say I want to determine how much money I should spend on a towy in order to ensure I'm getting a good value for the price.

Firstly, we need to determine what a "good value" is. Personally, if I'm paying more than one dollar per day that the toy functions, then it's not worth it. There's a start. I also don't want to always be going to the pet store, so if the toy can last at least five days, that would be optimal.

Secondly, I need to know how quickly she breaks her toys; She really likes breaking the big ones, those can be gone in a matter of a couple days, while some of the smaller ones (like those plushy mice with the tails that attack to a string) can last for weeks or even months. I estimate that for a toy weighing 100 grams, it takes her 5 days to destroy. Let's make a simple pseudo-model, using a simple systems of equations:

$$x = \frac{c}{d}$$

$$d = \frac{5}{s}$$

In this equation, if $x < 1$ and $d \geq 5$, I will purchase the toy, otherwise, I will not. Let's define the variables.

- x: The cost/durability ratio; and one of the main things which determines whether to purchase a given toy for her.

- c: The cost, in dollars/pounds/euros (your pick, it doesn't change the maths) of the toy.

- d: The durability of the toy, how long it will take before my cat decides it's time to tear the thing apart, measured in days. This is linked to the toy's size.

- s: The size of the toy, in units of 100 grams (the approximate weight of 100 paper clips).

Now it's simply a matter of plugging in the numbers for any given toy; let's say I want to purchase her a new string/ribbon toy. It is about 50 grams in weight and costs 3.50 of whatever currency I'm using in this book. So I go to the store with my calculator, and I punch in the numbers:

$$d = \frac{5}{0.5} = 10$$

$$x = \frac{3.5}{10} = 0.35$$

Since d>5 and x<1, I will be purchasing this toy; it meets both of my requirements. And just like that, you've helped me select a toy for my cat… I mean, you've learned how to do mathematical

modelling of simple, 1-dimensional equations.

Chapter 5: Logic Gates, Code, and Modelling

So now we understand some of the basics of mathematical modelling, numbers, and operations, we are ready to tackle the use of computer programming in order to solve mathematical concepts. This is where we pull everything together; and we will do that with computers.

Remember binary? Don't worry, we won't get into the direct inner workings of binary in a computer just yet, but this gives us an amazing opportunity to learn how to do mathematical modelling with Python. Instead, we will be having another look at how we can use binary values; to denote truth and falsehood.

Remember our previous chapter's algorithm; we're going to add one more factor into it: the size, in inches, of the toy. Let's rewrite our algorithm to incorporate this new factor (of course, this is all hypothetical):

$$x = \frac{c}{d}$$

$$d = \left(\frac{5}{s^{r}/100}\right)$$

The new variable, "r", is the size of the toy from its furthest two points, in inches. Once more, if d≥5 and x<1, we will purchase the toy. We can express this condition as the following:

$$d \geq 5 \,\&\, x \leq 1$$

`The ampersand ("&") simply means and, and denotes an AND gate. We'll have a closer look at other gates later in this chapter, but for now, know that "&" simply takes two inputs, and if both of them are true, it returns true. Sometimes, it's written as a carrot ("^"), and other times in diagrams, it's a symbol with one rounded edge, you'll know it when you see it (especially since in this book we'll write AND in the center of and symbols). Otherwise, it returns false. Let's draw a truth table to examine how this works:

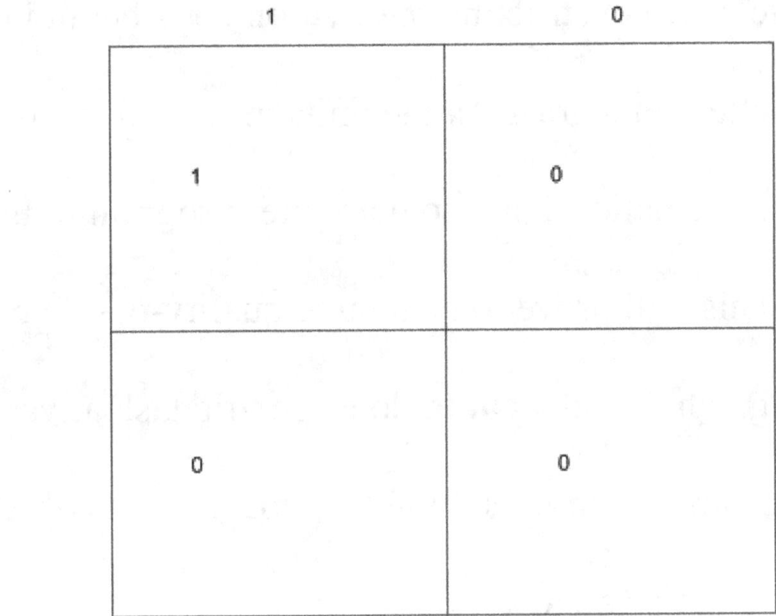

AND

In later chapters, we will examine how logic gates look when put into circuitry, but for now, this simple logic table will suffice. For now, we should have a look at writing the code in python.

It's also worth noting that, for simplicity's sake, I've redefined "s" to simply be the weight in grams, and as such the denominator, to compensate, has been divided by 100. This allows us to perform exponent operations on the gram amount without having to deal with the effect it has on the 100. This is really moreso to make it make more sense when we're deconstructing the python code behind it, and it also allows us to better get a grasp on the units, now that we're no longer using solely mathematics, but also computer programming, in solving our problem. This will prove to be a huge quality-of-life boost.

In coding, when you want to do a specific task at various parts, or you want to organise your code based on specific sets of instructions, you might use what's called a function. In our case, we will write a function called "catToyBuy", which takes our s, r, and c values as parameters. These will correspond to arguments which will be given to it by us through input, more on that later. In the meantime, let's look at the code for the function:

```python
def catToyBuy(s,r,c):
    d=5/((s**r/100))
    x=c/d
    if x<1 and d>= 5:
        return True
    else:
        return False
```

This code is the full code for our function; I kid you not, it is less than 8 lines. But first, it's worth trying to understand what a programming language even is. For that, I'm gonna use possibly the most overused analogy, which is basically baked into the term "programming language"

I have a question for you: Can you understand the following?

"Η γάτα είναι ένα μικρό ζώο που ζει σε σπίτια ή στη φύση και τρώει ποντίκια. Όπως οι σκύλοι, έτσι και οι γάτες είναι θηλαστικά."

Unless you speak Greek, you may be struggling to even read the passage; this is how a computer would think about our natural, human language. To the computer, the statement *"The cat is a small animal that lives in houses or in nature and eats mice. Like dogs, cats are mammals."* means absolutely nothing; it might as well be random letters. So to remedy this, we create programming languages with which we can communicate to computers.

Anyways, with that explanation out of the way, back to the code: Python is just about the closest we can get to reasonably having a computer be able to understand human language in a programming context, outside of large language-learning models like ChatGPT, at least for now, anyways. Because of this, Python lets us interact with a computer close enough to the way we talk that it's somewhat beginner-friendly.

Let's go line by line: firstly, we define what "catToyBuy" is; or more specifically, we declare the function and any parameters the function might have. Then, we perform our equations to get our "x" and "d" values, and if they meet our conditions, we return "True", otherwise we return "False". We use an AND gate here to help us determine if both of our conditions have been met.

Let's look at the rest of the code now:

```python
while True:
    s = float(input("How much in grams does your toy weigh? >>"))
    r = float(input("How big in inches is the toy? >>"))
    c = float(input("How much does the toy cost in dollars? >>"))
    result = catToyBuy(s,r,c)
    if result == True:
        print("It's worth the buy")
    else:
        print("It's not worth it")
```

This is also very simple code in general: We start a while loop, which will go on forever, since its condition is True. We then define three variables, each float values given to us from input (numbers capable of being decimals), and we use those variables as arguments to get our result from catToyBuy, saved to a variable named "result". Finally, if the returned value (stored in "result") is true, we print "It's worth the buy" to the console, and otherwise, we print "it's not worth the buy". Congratulations: You just wrote your first algorithm in Python!

Chapter 6: Logic Gates and Circuits

From code to circuits, we really are making quite the leap, aren't
we? But even then, we will still be working with logic gates. First, we
have to explain a few different components:

- Breadboard: A circuitboard in which pieces can easily be inserted
  and removed

- Transistor: A three-pin gate, which works as a switch, with two
  inputs and one output. We use NPN-BJT transistors.

- Diode: For our purpose, a diode is a one-way gate.

- Resistor: As the name implies, the resistor applies electrical
  resistance. Resistance is measured in ohms, or Ω. This resistance is
  described by Ohms law: $Resistance = \frac{Voltage}{Current}$, or $R = \frac{V}{I}$.
  Different combinations of colour bands on the resistors equate to
  different amounts of resistance.

- Push button: Literally just a button.

With any circuit, you need power and ground. The power flows through the circuit until reaching ground, and most circuit batteries have a ground pin built in, providing all your power flow needs in a single component.

Now, let's talk about the NOT gate; the simplest of the logic gates to understand. If you turn on the input, the output turns off, and vice versa.

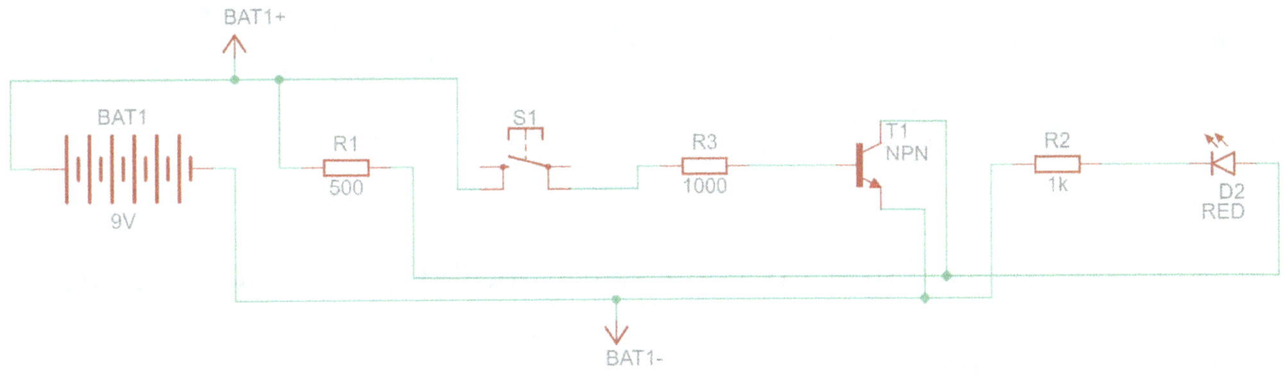

As you can see, trying to draw circuit diagrams using a virtual programme can be a hassle, but also here we have a NOT gate, using a transistor. Note that the resistor entering the collector is of higher resistance than that entering the base; this is intentional, as this way the circuit will actually work. The actual not gate, hooked up to nothing else looks like the following:2

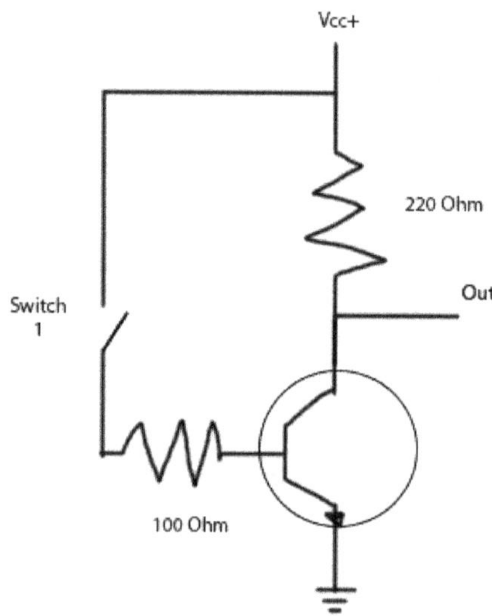

We can add this schematic to our toolbox; this will come back later. Let's have a look.

The whole circuit revolves around the transistor: From the constant voltage, we sent a resisted electrical current into the collector, with an offshoot before the collector being our output. Then, into the base, we sent a resisted electrical current controlled by a switch. Finally, at the emitter, there is ground. When the input switch is turned on, this causes the collector current to flow, and the voltage to drop near the collector current; thus preventing any high amounts of electricity from flowing through the output. When the collector is off, this is not the case.

Our next gate is the AND gate; we've already discussed how this one works from a logic perspective, so I'll just jump straight into the schematic.

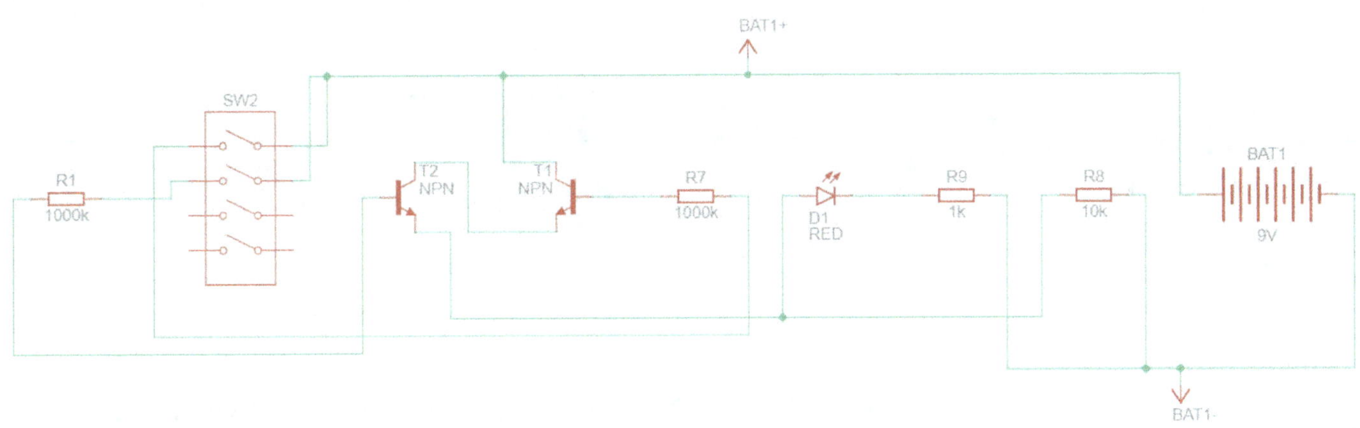

I used a different component for the input for this one; just so that I could more easily manage the input. Notice how one transistor outputs its emitter to the other's connector; this is what separates transistor AND

and OR gates. There isn't much else to say than that, but let's stuff this schematic into our toolbelt as well.

Time for OR: This gate activates as long as one of the inputs is on; meaning if the input is 10, 01, or 11, then it will activate. Behold, schematics!

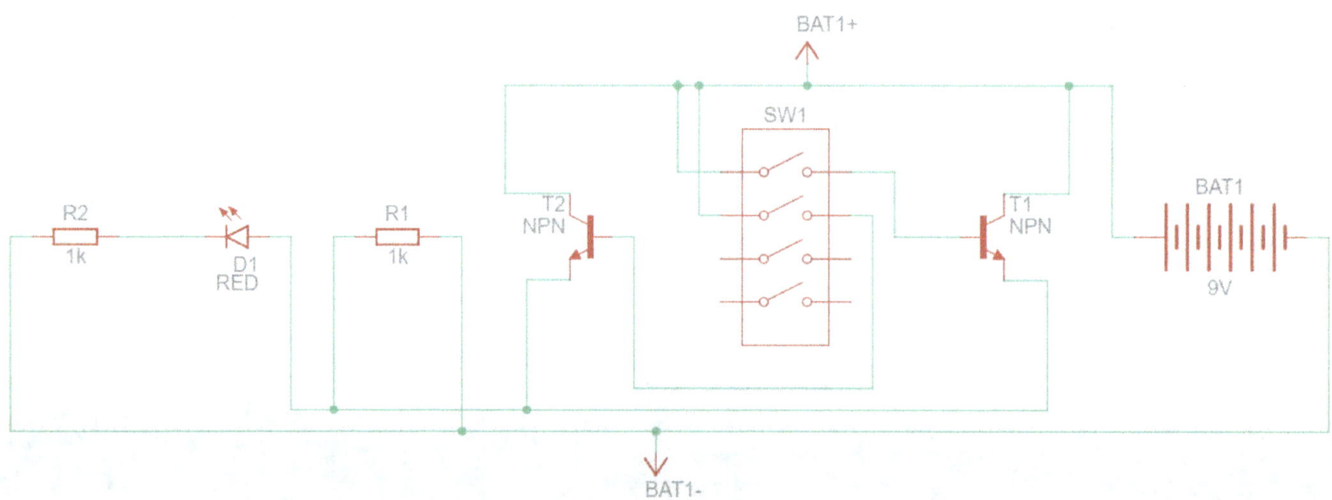

This time, the transistors are parallel, and do not rely on each other. This is arguably far more complicated than it needs to be though, since it's far easier to make one with diodes.

XOR, or exclusive OR, is like an OR gate, but it does not activate if both are on; only when exactly one input is on will it activate. Luckily, instead of a circuit diagram, I can do a logic diagram for this one, which

will be much easier to read, and a lot easier to make as well.

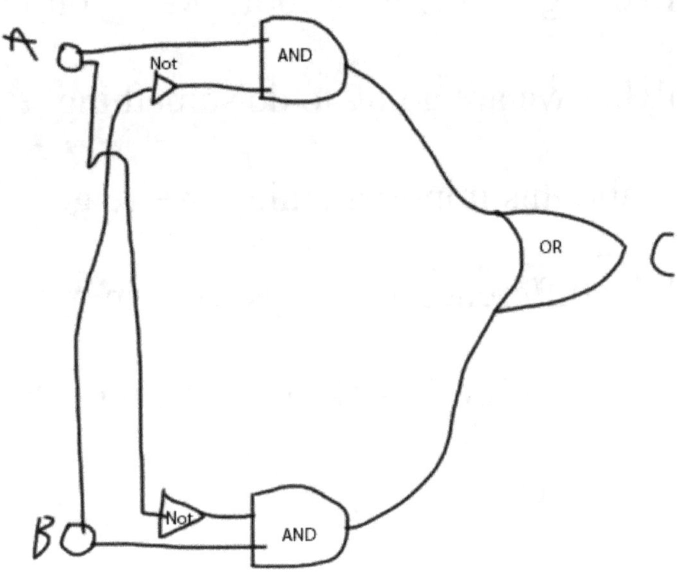

In this chapter, we examined basic logic gates. In the next chapter,

we will expand on our understanding of logic and how it applies to

mathematics, and ultimately, how we can use our knowledge of these

things to compute basic logic operations on bytes, both in programming

and flowcharts for manual maths.

Chapter 7: Logic to Solve a Problem

Would you like to design something cool? Well, you're in luck! Okay, it's not that cool, but we are going to do something important. In order to understand this important thing, we're gonna start with developing an idea of how different logic gates work in byte architecture, instead of just with bits. Let's look at some simple byte architecture which performs AND onto two bytes

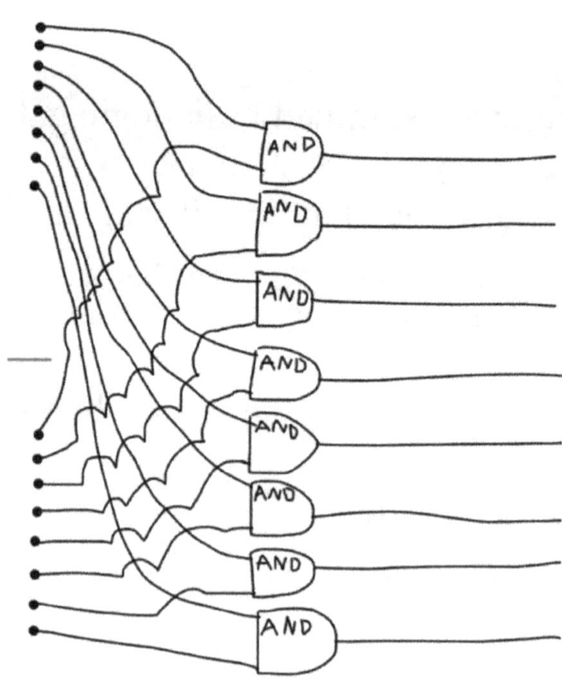

This logic schematic performs AND on two bytes. We can do similar things with OR, XOR, and even NOR and NAND gates (NOT + OR and NOT + AND).

We can use these logic gates to do some basic addition: Consider the fact that whenever we add two digits together in binary, we're already mentally doing a series of XOR gates, AND gates, and OR gates. Let me demonstrate, by asking you to add the two following numbers:

$$00101101_2 + 00010110_2$$

This is done by looking at each digit, performing an XOR, and then carrying the one if necessary to the next digit, bringing us to the answer of 01000011_2. We can simulate that process with the following Python programme, on the following page because it won't fit on this one:

```python
def XOR(A, B):
    if ((A == "1" and not B == "1") or (A == "0" and not B == "0")):
        return True
    else:
        return False

while True:
    A = input("byte 1 >>")
    B = input("byte 2 >>")
    C = ""
    Carry = False
    A = "00000000" + A
    B = "00000000" + B
    for i in range (len(A)-1, -1, -1):
        if (XOR(A[i],B[i]) and not Carry):
            C = "1" + C
            Carry = False
        elif (A[i] == "1" and B[i] == "1" and not Carry):
            C = "0" + C
            Carry = True
        elif(A[i] == "0" and B[i] == "0" and not Carry):
            C = "0" + C
            Carry = False
        elif(XOR(A[i],B[i])and Carry):
            C = "0" + C
            Carry = True
        elif (A[i] == "1" and B[i] == "1" and Carry):
            C = "1" + C
            Carry = True
        elif(A[i] == "0" and B[i] == "0" and Carry):
            C = "1" + C
            Carry = False
    print (C)
```

Here we can see the basic logic; I avoided converting these numbers into decimal (obviously), and because of that I had to do XOR gates manually. Let's observe how it works:

First, we define a function for XOR gates. If we were using integer or boolean values (something we'll later), then we wouldn't need to do this part, but since I kept things in strings to make the explanation part easier, we had to create our own XOR function. Next, our while

True loop runs, giving us our main loop of the programme; here I declare a few variables. A and B are both given by input, and C and Carry are declared with values. Now, it's time for me to explain booleans.

Booleans are simply one bit; a true or false. This also corresponds to a 1 or 0, and thus can be used in very basic calculations. Simply put, Booleans are a single one, or a single zero. Booleans are incredibly important for logic, as we'll see soon.

From here, we start a for loop; and I just realised I never explained the two different types of loop.

- For: A "for" loop is a loop which iterates a certain number of times: It does the same few operations over and over again until it it's done with it's count

- While: A "while" loop, on the other hand, iterates over and over while a certain condition is true, and if that condition is made false, it ends.

After that, the logic gates come in, and depending on the inputs and carry, it adds the next digit to the front of the new value, and updates the carry respectively.

I love this particular project because it allows us to examine in detail a process for doing something which, to most, comes naturally; it examines the underlying logic at how we arrive at conclusions without simply being another instance of blindly using the same methodology again.

Let's go ahead and test it out with our example from before, which, if you will recall, is $00101101_2 + 00010110_2$. Now personally, I ran this through the debugger, but as long as we know that it produces the correct result of 01000011_2, we should be fine.

Running the programme, it goes through, does exactly as we tell it to, and then returns our result, 0000000001000011. Of course, the computer adds a second byte to the beginning for digit padding, so that if our answer happened to be above 255, it could still represent it, but in our case it didn't need to do that. That being said, the answer is correct, as once you remove eight zeros from the beginning, you get 01000011_2, which is, of course, the exact answer which we were expecting to get, and the correct answer for this addition.

Chapter 8: Gravitational Forces

Gravity; it's the thing that keeps you on the ground, the planets orbiting the stars, and our sun around the Milky Way. But, as we will come to learn, this mysterious force is actually pretty simple to understand, and really, not even that complicated.

Firstly, let's define force in terms of mathematics. Force can be defined as the following:

$$F = ma$$

Force is mass times acceleration, or, in other words, if you multiply the amount of stuff in something and the change in directional speed that something is experiencing, you will get the force acted upon it. Gravity has a special formula for calculating it's force that an object exerts on another object, which is the following:

$$F = \frac{GM1M2}{D^2}$$

This is best summarised as being the gravitational constant "G", which is about $6.67e^{-11}$, or $6.67*\frac{1}{100,000,000,000}$, times the mass of both objects, all divided by the distance between those objects squared. Now, we can combine these two formulas to get the following:

$$M1a = \frac{GM1M2}{D^2}$$

$$M2a = \frac{GM1M2}{D^2}$$

If you remember the chapter on basic algebra and solving equations, you may already know what I'm gonna do here, but I want to find out the acceleration between these two objects. Not to worry, since we can cancel out the masses on both sides:

$$a = \frac{GM2}{D^2}$$

$$a = \frac{GM2}{D^2}$$

Following along still? Because we will be using this information later in programming a gravitational acceleration calculator. In the meantime, let's look at an example problem.

My eternally adventurous cat has decided to venture from my basement and living room to the vast expanses of outer space. No clue why, but just say, hypothetically, she did. Also, she chose to take with her my bowl of popcorn; a delicious snack, which no one can convince me to stop eating.

Let's assume that for her breed, she's pretty healthy (she is, and technically it's my sister's cat, so I don't have her weight records immediately available, but she certainly acts healthy and a lot more active than I ever could be). That puts her somewhere around 4.5 kg, given that she's a Manx. My bowl of popcorn, however, is much lighter, say 60 grams. I don't know the actual amount, I'm just pulling a number out of thin air.

Let's do our calculations! Have a look:

$$F = \frac{GM1M2}{D^2}$$

Now, I'm going to assume my popcorn is one and a half metres away from my mischievous cat; so we can fill in all of the variables.

$$F = \frac{(4.5)(0.06)G}{1.5^2} \Rightarrow$$

$$\frac{(4.5)(0.06)G}{(1.5)(1.5)} \Rightarrow 2(0.06)G \Rightarrow 0.12G \Rightarrow 8.004 * 10^{-12}$$

That's… not very much force, which is to be expected, because gravity is extremely weak, and we only notice it because space objects are super massive. Anyways, with this in mind, we can find the acceleration of both my cat and my popcorn now:

$$8.004 * 10^{-12} = 4.5a \Rightarrow 1.778 * 10^{-12} = a$$

$$8.004 * 10^{-12} = 0.06a \Rightarrow 1.334 * 10^{-10} = a$$

So, now we know my cat is being hurtled at my snack at insanely slow speeds. To put that into context, at the speed of the popcorn, if I were to start walking at the beginning of the universe and travel constantly from where I am on earth right now with no breaks, I would not have made it a sixth the way to the moon yet; yeah, that's pretty slow. I think I'm safe from having my cat steal my snacks anytime soon.

In the next chapter, we'll talk about designing a computer programme to do these calculations for us, but for now, just marvel at the sheer expanse and scale of our universe, if something even as relatively small as our planet can hold us to it consistently.

Chapter 9: Gravitational Maths, and More Python

Maths, as you can imagine, even as it relates to gravity, can easily be programmed into a computer to be calculated without us even having to pick up a calculator; This is actually really simple to do, it will actually take us less time than the binary adder.

Using the knowledge and the mathematics we learned in the previous chapter, we can very easily write our programme. However, we should introduce one more data type: lists.

Lists simply contain a list of various different values; they can be the same type or different types, but regardless of what they are, they are all part of the list's index(starting at index "0"); think of the list as a library, and all of the individual index values as placements of books; if you go to a specific index value and search it, you'll find the book you want; but there's no sense checking the CompSci section for a book about Hockey!

Alright, so now that we've established the basics of what a list is (which is important because our function will be returning a list as output), we can examine the idea of writing code to understand gravity.

Have a look at the following bit of code:

```python
def calculator(M1,M2,D):
    G=6.67*(10**-11)
    F=G*M1*M2/(D**2)
    A1=F/M1
    A2=F/M2
    return [F,A1,A2]

while True:
    M1 = float(input("Mass 1(kg): "))
    M2 = float(input("Mass 2(kg): "))
    D = float(input("Distance(m): "))
    results = calculator(M1,M2,D)
    print("Force: " + str(results[0]))
    print("Acceleration 1: " + str(results[1]))
    print("Acceleration 2: " + str(results[2]))
```

By now you can probably break down the entirety of this code yourself, but it *is* my job to do so, so I'll just quickly go over the calculator line by line.

- "def calculator(M1, M2, D):" simply declares a function "calculator", with parameters M1, M2, and D, which we will use to calculate everything else.

- "G=6.67*(10**-11)" defines the gravitational constant, as described in the previous chapter. You know, $6.67*10^{-11}$, the thing that made gravity so weak? Yeah, that thing.

- "F=G*M1*M2/(D**2)" is simply the formula to get our force ("F") given our inputs

- "A1=F/M1" get's the acceleration of object 1

- "A2=F/M2" is… you get it.

- "return [F,A1,A2]" simply makes the function output F, A1, and A2 as a list

Oh, also, *float* and *str* are both functions used to convert a value to a float or string respectively, Float values are numbers with a decimal, while strings are text.

Now that we have that programme, feel free to play around with it; plug in numbers from various celestial bodies (which you can look up on your preferred search engine), or just plug in numbers of things around your house and look at how small the numbers can be.

While you're doing that, I'd recommend that you also take the time to play around with lists in a separate programme, or maybe just a shell (which… I still haven't explained, but it's basically an environment where you write your code line by line and execute it as you run it). Anyways, now that we've integrated gravity into our calculations with Python, it might be worth thinking of other equations you can try to write code for.

Chapter 10: Classes

Python is object-oriented; this means that, in Python, you can create "objects", which function sort of like real objects; they have properties. You may remember that before we were breaking up objects into their most simple mathematical concepts: For instance, my cat's weight was used as a standalone variable.

When we make an object in python, this object is called a "class." A class is capable of holding various values for an object, which you declare in a function called "__init__". This initiation function is where you declare the parameters of the class.

Let's have a quick revisit at what we did in the last chapter: We learned how to write a programme to simulate gravity. This, however, had some shortcomings: Firstly (and at no fault of non-object programming), our programme assumed we knew the distance from one object to the other; I'd like to swap that for a 3-dimensional coordinate positioning system, since that does not require us to know distance, but simply the points our objects are at. Secondly, I would like to save the values of each object into a class; the name, mass, and coordinates of each object stored in a set of classes would be a *lot* easier in many different ways; mainly, we wouldn't need to repeat putting in the same value; we could just use the name as an identifier and then call the attributes as needed. Let's have a look at classes.

Classes, in Python, are surprisingly easy to write. You really just need to declare the class with the "class" keyword. Have a look at this example:

```python
class Person:
    def __init__(self, name, age, height):
        self.name = name
        self.age = age
        self.height = height
    def ageself(self):
        self.age = self.age + 1

p1 = Person("John Doe", 19,67)
print(p1.name + " is " + str(p1.age) + " years old.")
p1.ageself()
print(p1.name + " just turned " + str(p1.age) + "!!! Happy birthday, " + p1.name + "!")
```

A class can hold both functions and attributes; and in our example here, the attributes are "name", "age", and "height", and we have a function "ageself" which ages the person.

Now, let's see some actual code that puts this into action! Our scenario? The same as the last chapter; we want to calculate gravitational acceleration and force. I've already written the code, and I encourage you to look at it and try to analyse it yourself *before* looking at the explanation.

```python
class celestialBody:
  def __init__(self, name, mass, position):
    self.name = name
    self.mass = mass
    self.position = position

def getCommand(command):
  command = command.lower()
  if command == "create body":
    return 0
  if command == "find values":
    return 1
  if command == "list bodies":
    return 2

celestialBodies = []
while True:
  command = input("What would you like to do?")
  commanded = getCommand(command)
  if commanded == 0:
    name = input("Name: ")
    mass = input("Mass: ")
    x = input("X Coord: ")
    y = input("Y Coord: ")
    z = input("Z Coord: ")
    name = name.capitalize()
    mass = float(mass)
    x = float(x)
    y = float(y)
    z = float(z)
    celestialBodies = celestialBodies + [celestialBody(name,mass,[x,y,z])]
  if commanded == 1:
    name1 = input("Name 1: ")
    name2 = input("Name 2: ")
    name1 = name1.capitalize()
    name2 = name2.capitalize()
    for i in range(len(celestialBodies)):
      if celestialBodies[i].name == name1:
        index1 = i
      if celestialBodies[i].name == name2:
        index2 = i
    D = (((celestialBodies[index1].position[0]-celestialBodies[index2].position[0])**2
        +((celestialBodies[index1].position[1]-celestialBodies[index2].position[1])**2)
        +((celestialBodies[index1].position[2]-celestialBodies[index2].position[2])**2))**(1/2)
    G = 6.67*(10**-11)
    F = G*celestialBodies[index1].mass*celestialBodies[index2].mass/D**2
    A1 = F/celestialBodies[index1].mass
    A2 = F/celestialBodies[index2].mass
    print("D = "+str(D)+"\nF = " + str(F) +"\nA1 = " + str(A1) + "\nA2 = " + str(A2))
  if commanded == 2:
    for i in range(len(celestialBodies)):
      print(celestialBodies[i].name,celestialBodies[i].mass,celestialBodies[i].position,)
```

It may be a bit small to read, given how long it is, but let's analyse the code.

The first portion is the class; it defines what a celestial body is, including declaring its name, mass, and position in space. These values are initialised in the "__init__" function (which basically all useful classes have). I've also defined a function for handling commands, but that one is easy enough to understand just by reading it.

Once we have our command, we can either make a celestial body, which will add it to the list of objects, or we can calculate the force, acceleration, and distance between two objects exerting gravity on one another, and, of course, we can list our celestial bodies.

All this ties together the last few chapters, but this isn't the only use of these programming components: In the next chapter, we'll explore how these operations can be used to write code to do bitwise functions.

Chapter 11: Bitwise and Classes

Now that we officially know basic logic operations, and how to write classes, it's time to pull all of that together. That's right; we're using classes to assist us in our bitwise operations now.

To start off, I'd like to remind you of all the logic gates we've looked at: AND, OR, NOT, and XOR. AND refers to the gate which looks at two values, and if both of them are true, it will return true. OR looks to see if at least one is true, XOR looks to see if exactly one is true, and NOT takes one input and inverts it.

These bitwise functions can also be applied to bytes; after all, a byte is just a series of 8 bits. Therefore, with a little bit of looping, we can easily write a programme capable of doing all of these.

If we can write a programme capable of doing all of these, and a programme capable of adding two binary bytes together (as we already have), we would have the beginnings of what could eventually become a turing machine if worked on hard enough.

To summarise, a Turing machine is simply a computer; it can take inputs and return appropriate outputs (something which we will discuss in a later chapter) by manipulating bytes of information. In our programme, we will work to make a programme capable of performing bitwise functions.

Luckily, I've already written the code. Feel free to look at it and figure out how it works, but I will also break it down step by step:

```python
#Class declaration
class ByteIndexer:
    def __init__(self, *bytesls):
        self.bytesls = bytesls
```

First, as seen above, we have our class declaration, fit with our __init__ function. It takes a variable amount of bytes in its initiation (that's what the asterisk means). Of course, this means that we can feed it a bunch of bytes immediately if we really want to, but we also have the means to add more bytes to it later (more on that when we reach that

point).

```python
#This is mainly for the addbytes function, to allow us to get an XOR operation for one bit
def XORbit(self, A, B):
    if ((A == "1" and not B == "1") or (A == "0" and not B == "0")):
        return True
    else:
        return False
#NOT gate for one byte
def NOT(self, A):
    A = self.bytesls[int(A)]
    out = ""
    for i in range(len(A)):
        if(A[i]=="1"):
            out = out + "0"
        else:
            out = out + "1"
    return out
#AND gate for two bytes
def AND(self, A, B):
    A = self.bytesls[int(A)]
    B = self.bytesls[int(B)]
    out = ""
    for i in range(len(A)):
        if(A[i]=="1" and B[i] == "1"):
            out = out + "1"
        else:
            out = out + "0"
    return out
#OR gate for two bites
def OR(self, A, B):
    A = self.bytesls[int(A)]
    B = self.bytesls[int(B)]
    out = ""
    for i in range(len(A)):
        if(A[i]=="1" or B[i] == "1"):
            out = out + "1"
        else:
            out = out + "0"
    return out
#XOR gate for two bites
def XOR(self, A, B):
    A = self.bytesls[int(A)]
    B = self.bytesls[int(B)]
    out = ""
    for i in range(len(A)):
        if((A[i] == "1" and not B[i] == "1") or (A[i] == "0" and not B[i] == "0")):
            out = out + "1"
        else:
            out = out + "0"
    return out
```

After that we have the code for various logic operations: NOT, AND, OR, and XOR.

I also included a separate XOR operation for a single bit, that way we can keep the functionality of our adder the way we had it; which brings me to something important. As a computer programmer, it's totally fine to reuse code over and over again from previous things you've made (unless you're in a class setting or a work setting and the teacher/boss says otherwise). As long as you wrote the original code yourself, no one is gonna get mad that you've used elements of something you've already made. Being resourceful with your code is a good trait for a programmer, after all.

```python
#Adds two bytes together
def addbytes(self,A,B):
    C = ""
    Carry = False
    A = "00000000" + str(self.bytesls[int(A)])
    B = "00000000" + str(self.bytesls[int(B)])
    for i in range (len(A)-1, -1, -1):
        if (self.XORbit(A[i],B[i]) and not Carry):
            C = "1" + C
            Carry = False
        elif (A[i] == "1" and B[i] == "1" and not Carry):
            C = "0" + C
            Carry = True
        elif(A[i] == "0" and B[i] == "0" and not Carry):
            C = "0" + C
            Carry = False
        elif(self.XORbit(A[i],B[i]) and Carry):
            C = "0" + C
            Carry = True
        elif (A[i] == "1" and B[i] == "1" and Carry):
            C = "1" + C
            Carry = True
        elif(A[i] == "0" and B[i] == "0" and Carry):
            C = "1" + C
            Carry = False
    return (C[0:8],C[8::])
#Outputs any number of bytes
def outputByte(self,*A):
    outbytes = []
    for i in A:
        outbytes =outbytes + [self.bytesls[i]]
    print(outbytes)
    #Creates new bytes
def newBytes(self,*A):
    for i in A:
        self.bytesls = self.bytesls + i
```

Above is the next portion of the code for the ByteIndexer class. It focuses on the remaining operations that I wanted to include in this version, including operations for addition, the ability to print bytes to the console, and the ability to add new bytes to our existing list of bytes.

```python
BI = ByteIndexer("10110110","01101101","00010110")
#Here, you can write your own code to test out this class
BI.newBytes(BI.addbytes(0,1))
BI.newBytes(BI.addbytes(0,4))
BI.outputByte(slice(0,len(BI.bytesls)))
print(BI.NOT(4))
print(BI.AND(3,6))
print(BI.OR(3,6))
print(BI.XOR(3,6))
```

This portion of the code is literally just our test cases; we create an instance of ByteIndexer called BI, and we force it to do some of its operations.

When we combine all of this, we have a machine that can very easily manipulate bytes, including adding new bytes to itself and performing logic operations on bytes. I'd encourage you, the reader, to write your own supplementary functions to go along with this programme; can you figure out how to write code for a function that sets a specific byte in the byte index to a certain value? Go ahead, try it! Of course, I already have myself, but I'm saving it for later chapters, where we may or may not try to design a simplified version of a turing machine. Firstly, however, let's take a bit of a break from programming to get back into numbers.

Chapter 12: Calculus and Changes in Change

If you took calculus in highschool or college, you might remember the concept of Derivatives quite well, maybe you even remember how to differentiate functions. Luckily for us, polynomial differentiation is pretty easy; for every term in a polynomial, perform the following formula (one might call it an algorithm, purely to make this chapter feel more connected to the previous 7):

$$\frac{d}{dx}(x^a) = ax^{a-1}$$

For instance, x^2 would differentiate to 2x. Let's look at a longer example, shall we?

$$\frac{d}{dx}(x^3 - 3x^2 + 5x - 2)$$

Here, we just go term by term, differentiating the polynomial one term at a time. Here's what we get in the end:

$$\frac{d}{dx}(x^3 - 3x^2 + 5x - 2) \Rightarrow 3x^2 - 6x + 5$$

Oh, I should have mentioned; the derivative of a constant is always 0. Anyways, now that we know how to find the derivative of a polynomial, you may be asking yourself a few questions, such as "what did we even calculate?" Well, the derivative tells us the slope of a function at any given point; so I'd say that's pretty neat, especially given the things we can do with that; but before we continue, let's examine the generalised power function rule of differentiation:

$$\frac{d}{dx}(f(x)^n) = nf(x)^{n-1}f'(x)$$

If we think back to the rule we just learned about polynomials, this makes a lot of sense; since the derivative of x is 1, the derivative of x^2 is $2(x^{2-1})(1)$, or $2x$. However, this generalised rule also

allows us to evaluate derivatives of functions such as the following:

$$\frac{d}{dx}(x^2 - 4x + 4)^2$$

To evaluate such a function, we simply have to plug in the

right things into our rule: doing so yields us this:

$$\frac{d}{dx}(x^2 - 4x + 4)^2 = 2((x^2 - 4x + 4)^{2-1}\frac{d}{dx}(x^2 - 4x + 4)$$

Then if we solve the derivative of x^2-4x+4, we get 2x-4,

allowing us to finish solving our derivative equation:

$$2(x^2 - 4x + 4)(2x - 4) \Rightarrow 4x^3 - 24x^2 + 48x - 32$$

Simple enough, but still wildly important to our

understanding of mathematics; an achievement which allows us to

calculate some important things, such as a lot of physics, as differential calculus is at the root of how kinematics works. For instance, if my acceleration is changing by 2t every second, then my velocity is changing by t^2 every second, since acceleration is just the first derivative of velocity.

Let's have a quick look at some graphs, shall we? A ball is accelerating from standstill at 3 metres per second (m/s) down a hill;

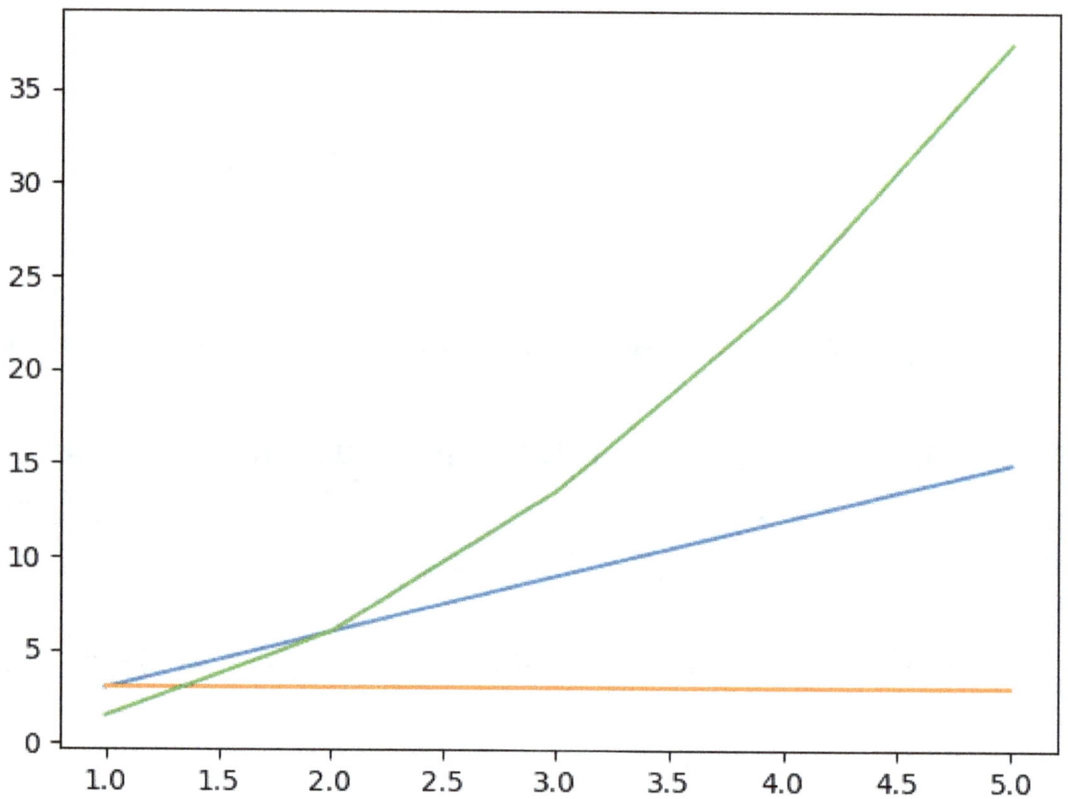

This graph illustrates the position, velocity, and acceleration of the supposed ball; notice how the ball's acceleration (orange) is a constant, the ball's velocity (blue) is a monomial of the first degree, and the position (green) is a monomial of degree 2 (hence why it curves upwards slightly).

But wait! You might be saying, *How did you get the values for the velocity and position?* Well, glad you asked, hypothetical reader! This is what's known as the antiderivative, or the function which a function is the derivative of.

Basically, since it's so easy to find the derivative of a polynomial/monomial, it takes very little effort to find the reverse of such a thing as well, hence the antiderivative. From there, the antiderivative of acceleration is velocity, and the antiderivative of velocity is position.

Chapter 13: Back to Class (and Files)

Now that we've taken a break, let's get back to class; Python classes. It's time to look back at our programme from Chapter 11. We're still not ready for a Turing machine, and it's likely we won't end up making one at all, but for now, we can at least incorporate file input.

```python
def getDecimal(binary):
  out = 0
  for i in range(0,len(binary)):
    if binary[i] == "1":
      out = out + 2**(len(binary)-(i+1))
  return out
def doFileInstructions(fileName,BI1):
  pointer = 0
  f = open(fileName+".txt", "r")
  for line in f:
    args = line.split()
    print( args)
    if args[0] == "NOT":
      BI1.override(pointer,BI.NOT(getDecimal(args[1])))
    elif args[0] == "AND":
      BI1.override(pointer,BI.AND(getDecimal(args[1]),getDecimal(args[2])))
    elif args[0] == "OR":
      BI1.override(pointer,BI.OR(getDecimal(args[1]),getDecimal(args[2])))
    elif args[0] == "XOR":
      BI1.override(pointer,BI.XOR(getDecimal(args[1]),getDecimal(args[2])))
    elif args[0] == "ADD":
      x=BI.addbytes(getDecimal(args[1]),getDecimal(args[2]))
      BI1.override(pointer,x[0])
      BI1.override(pointer+1,x[1])
    elif args[0] == "MAKE":
      BI1.newBytes(args[1])
    elif args[0] == "PUSH":
      BI1.override(getDecimal(args[1]),args[2])
    elif args[0] == "POINTUP":
      pointer = pointer + 1
    elif args[0] == "POINTDWN":
      pointer = pointer - 1
  f.close()
  f=open(fileName+" OUT.txt","w+")
  for i in BI1.bytesls:
    f.write(str(i))
    f.write("\n")
  f.close()
  return BI1
BI = ByteIndexer("00000000")
#Here, you can write your own code to test out this class
inpt=input("Give me some input file>> ")
doFileInstructions(inpt, BI)
```

As you can see, our code has gotten a lot more complicated; however it's nothing that a little close reading can't decipher.

To start off, we've added a function outside of the ByteIndexer class, which allows us to convert binary numbers into decimals. This will allow us to better write instructions; using this function allows us to use our binary inputs as decimals for indexing lists. We do this with the same mechanism we would by hand.

Next is the big one; doFileInstructions. This function reads from a file (a new thing, which is done with the "open" function), and it reads from that file, manipulating an indexer using several commands. Each of these commands (or as they're technically called "keywords"), and most of the functions use a pointer to determine where to output.

Then, we write the output to a text file, and return the indexer as our output for functions.

So far this seems like simply a novelty, but it really does have an important lesson to teach us; this programme contains the basics which can, eventually become a programming language, and even with what we have now, our code could be considered, in a way, a very simple look into programming language development.

The keywords, in particular, took a while to write. ADD threw me fore a loop (pun, of course, intended) because it required a different set of instructions to make work the way I wanted it to. Now, let's have a simple look at some instructions for this programme. I wrote

some in a text file just to test if it worked, here's the input and output I
got, but I should also mention beforehand I changed the override
function a tiny bit

```python
def override(self, A,B):
    A=int(A)
    if len(self.bytesls) > A:
        self.bytesls[A] = B[-8::]
    else:
        self.newBytes(B)
```

All the new change does is allow us to use the override function to
append a new byte, if there is no longer any space to do so. This is really
important because it allows the "ADD" keyword to function properly
even if the pointer is at the end of the list of bytes. Anyways, let's have a
look at that input and output now:

MAKE 11010010	00000000
MAKE 00101101	11010010
MAKE 00000000	00101101
POINTUP	00000000

POINTUP POINTUP ADD 00000001 00000010	11111111

Let's look line-by-line. To start off, the ByteIndexer now starts with a default byte of "00000000_2", or "0", at the very beginning of the programme, to give the users something to work with and to ensure that the instance started will function properly. We append two bytes, one of which is the opposite; or the "NOT", of the other. We then add an empty byte and move the pointer up three, to bring it to this empty byte. We add byte 2 and 3 to make byte 4 and 5 reflect the addition; which, in this case, happens to equal 255, or 11111111_2. This functions exactly as expected; and that's the thing with programming languages, they do exactly what you tell them to do. They assume nothing on your end, and they treat every instruction given completely literally, within the confines of their syntax. That's why we have to be very careful when writing code; it can very easily become tedious very quickly, and make the whole task of automating things or creating algorithmic solutions far more complicated when it fails to work the way you expect. Let's look at another set of instructions that results in the same output:

<table>
<tr><td>

MAKE 11010010
MAKE 00000000
POINTUP
POINTUP
NOT 00000001
MAKE 00000000
POINTUP
ADD 00000001 00000010

</td><td>

00000000
11010010
00101101
00000000
11111111

</td></tr>
</table>

Our new code, as you can probably assume, simply does the same thing, except instead of manually assigning the inverse of byte index number 1 to byte index number 2, we have the computer perform a NOT gate on byte index number 1. This means that if we change byte index number 1, the final two lines, where the two bytes are added, will not change.

Chapter 14: Loops, Loops, and More Loops

So, you've learned quite a bit about programming so far, but we really haven't touched on loops yet; which is a bit of an oversight on my part, because loops, along with conditionals, are amongst the most important parts of computer logic.

You've probably already seen loops in programming; especially if you've been reading this book cover-to-cover, since I use them all the time You've probably seen me use a *for* and *while* loop, and for this specific part of the book, I'll likely be using Java; firstly, I use Java for class so I already have that on my mind, but secondly, being able to transfer your logic from one programming language to another is a super useful skill.

For now, let's look at this in terms of dice; later on, we'll talk about finding the probability of dice by hand, but for now, let's focus on simply writing a programme in Java to do it for us; We could use maths, but since I'm saving the maths aspect for a future chapter, for now, we

will make our Java programme simulate rolling dice.

```java
import java.util.Scanner;
public class DiceProb{
    public static void main(String[] args){
        //We need a scanner in Java to take input
        Scanner in = new Scanner(System.in);
        //Let's declare the necessary variables
        int numOfDice = in.nextInt();
        int dice = in.nextInt();
        int sides = in.nextInt();
        int modifier = in.nextInt();
        int count = 0;
        /*
        For testing:
        System.out.println(numOfDice);
        System.out.println(dice);
        System.out.println(sides);
        System.out.println(modifier);
        */
        double average = 0;
        //Here's our loop; a do loop is like a while loop, except a do loop has a guarantee of executing at least once
        do{
            //We increase the cound
            count = count + 1;
            //A for loop is another type of loop
            //This loop differs from while loops, since this has a defined number of repetitions
            //In other words, with a for loop, we already know vaguely how many times it will repeat
            //With a while loop, we can be more uncertain
            //In this case, we know the loop will repeat once for every number between 0 and the number of dice per roll
            //In python, a for loop is done using different syntax; seen in previous chapters
            for(int i = 0; i < dice; i++){
                //This is the body of our loop
                //This is the code that repeats
                int added = 1+(int)Math.floor(Math.random() * (sides));
                average += added;
                //More stuff for testing
                /*System.out.println(added)*/;

            }
            //As this is still within our "do" loop, this will execute every time the do loop happens, at the very end
            average += modifier;
        }while (numOfDice > count); //This is the do loop's condition
        average = average / numOfDice;
        //And this is a formatted print statement
        System.out.printf("Average: %.4f",average);
    }
}
```

This is our programme; it's not exactly the most friendly looking

thing, I'm aware; but it allows us to examine how loops work. Loops

are, essentially, sections of code that you want to execute over and over

again. Imagine a loop like a song being played on repeat; until you do something to exit the repetition of the song, the song is gonna keep on playing.

As you can see, we've used two different types of loops here, one nested inside another. The do loop has a for loop inside of it. A "for" loop is most useful when we know how many repetitions we need to do, while "do" and "while" are much better for unspecified quantities of repetition.

Chapter 15: Probability and Mathematics

Probability is a very fun study of mathematics, and it's one we can easily examine both by hand, and using computers; but more specifically, in this chapter, we will be examining things by hand first.

Say, by chance, you're playing a game of Texas hold'em/poker for fun with three others (by the way, I am not endorsing gambling here, if any children are reading this book and have made it this far, hopefully you're smart enough to know that gambling is a losing game). Since you're not gambling money (because, as previously said, gambling isn't a good idea), you're betting sweets. You have a total of 55 sweets, as does everyone at the table. The pre-flop (first two face-down cards) have been dealt, and you need to decide how much to bet. Each player has already put in their first bet (5 sweets), leaving everyone with 50 (before you get on my case, I know ante-up isn't standard in Texas hold'em, just go with it for now).

Now, obviously, your goal is to come out of the betting with more glucose-packed sweets than you initially started with; so in order to do that, we need to predict what scenarios could happen from this point onwards. This is called the "expected value". We get this by multiplying the value of each outcome by the probability that it will happen;

Now that you've received your two cards, you know that you have the queen of spades and the two of spades. You're first up to make your bet; do you raise for fold?

Welp, let's have a look at what can happen this round, and for the sake of simplicity, I will not be including check as an option, since check basically copies the previous choice:

You (P1)	P2	P3	P4	Outcome
Fold	Fold	Fold	Raise	Loss
Fold	Raise	Fold	Fold	Loss
Fold	Fold	Raise	Fold	Loss
Raise	Fold	Fold	Fold	Win
Fold	Raise	Raise	Fold	Continue to next round
Raise	Fold	Fold	Raise	Continue to next round
Raise	Raise	Fold	Fold	Continue to next round

Raise	Fold	Raise	Fold	Continue to next round
Fold	Raise	Fold	Raise	Continue to next round
Fold	Fold	Raise	Raise	Continue to next round
Fold	Raise	Raise	Raise	Continue to next round
Raise	Raise	Raise	Raise	Continue to next round
Raise	Fold	Raise	Raise	Continue to next round
Raise	Raise	Fold	Raise	Continue to next round
Raise	Raise	Raise	Fold	Continue to next round
Fold	Fold	Fold	Fold	Continue to next round

Assume, for the sake of simplicity once again, that P1-P4 are playing randomly, and that if no one raises, P4 will definitely raise (thus winning the betting pot).

If you play "raise", you have about a 7/16 chance to continue to the next round, and you have a 1/16 chance to win outright. We can calculate the expected value this round as follows:

$$E(n) = \sum_{n=1}^{8} (r(n)p(n))$$

Where E(x) is our expected value, our r(x) is the value of the outcome, and p(x) is the probability of the outcome. For now, our value will be 1 for a win, -1 for a loss, and 0 for a "continue".

Value essentially means

If we raise, we can expect an outcome resembling the following:

$$0 * (7/16) + 1 * (1/16)$$

Using this, we see that our expected value is 1/16 of a sweet for raising this turn; that is to say, if we raise this turn, we have done the equivalent of winning 1/16 potential sweets.

Now let's look at what happens if we don't raise this turn. If we choose to fold, we have a 3/16 chance of someone else winning immediately, and a 5/16 chance of continuing to the next round. Using our formula from earlier, we get the following:

$$- 1 * (3/16) + 0 * (5/16)$$

I don't know about you, but I think I far prefer the idea of a 1/16 chance at earning over a 3/16 chance at losing, so we'll go ahead and bet; Rinse and repeat this process every round, and, assuming all else is done randomly, you can find your probability of winning the game for every round going forward.

Aside from it being obvious that I don't know the first thing about poker, I hope you've learned something from this; that knowing what your chances are with certain things can actually make decisions a lot easier and more obvious. For now however, let's explore programming, probability, dice, and how to ruin your DnD table.

Chapter 16: Dungeons and Computerised Statistics

DnD is a long-standing hobby for just about every friend group of slightly socially-awkward middle school through college students who only get the chance to hang out together once every second Wednesday between the hours of 6PM and 9PM, but aside from that, it has one very important aspect to our studies of maths, and specifically, statistics; the use of dice. I have a lot more experience with DnD than I ever will with poker, and dice is a lot easier of a computation, so hopefully this will go much better. Your party stops at a shop, and you see two weapons upon the wall; a Lance and a Maul. One of those does 1d12 damage, and one does 2d6 (side note; 1d12 just means one 12 sided die, and 2d6 are 2 six sided ones.) At a glance, it might not seem like there's much of a difference; but let's do the maths.

We can modify our formula from before, to instead reflect our new range.

$$E(n) \; = \; \sum_{n=2}^{12} (r(n)p(n))$$

Let's make a few tables. These tables will not be of wood nor stone, but rather, they're just tables of numbers. First, let's get a table for all of the outcomes that can occur from a roll of 2d6

--	1	2	3	4	5	6
1	2	3	4	5	6	7
2	3	4	5	6	7	8
3	4	5	6	7	8	9
4	5	6	7	8	9	10
5	6	7	8	9	10	11
6	7	8	9	10	11	12

Next, count the amount of times each number appears in a white cell; 2 and 12 appear once each, 3 and 11 appear twice, and so on and so fourth. Let's organise these into a new table, with the probability of each outcome and the outcome itself. While we're at it, we'll also add in the product of probability and value.

Number	Probability	Number * Probability
2	1/36	1/18
3	2/36=1/18	3/18 = 1/6
4	3/36=1/12	4/12 = 1/3
5	4/36 = 1/9	5/9
6	5/36	30/36 = 5/6
7	6/36 = 1/6	7/6
8	5/36	40/36 = 10/9
9	4/36 = 1/9	9/9 = 1
10	3/36=1/12	10/12 = 5/6
11	2/36=1/18	11/18
12	1/36	12/36 = 1/3

Now we should add those all up. When we do, we get 7; Interesting, isn't it? Well, that's just one side of the equation (or rather inequality, in mathematical terms)

Now, the other one is a lot simpler, and we can even skip to the second table

Number	Probability	Number * Probability
1	1/12	1/12
2	1/12	2/12
3	1/12	3/12
4	1/12	4/12
5	1/12	5/12
6	1/12	6/12
7	1/12	7/12
8	1/12	8/12
9	1/12	9/12
10	1/12	10/12
11	1/12	11/12
12	1/12	12/12

This yields the answer of 6.5; a number which, as you may have expected, is less than 7. So, it is mathematically better to use 2d6 than it is to use 1d12; Since on average, you're getting 0.5 higher on 2d6.

Okay, that's all well and good, but I promised in this chapter we'd go over how to actually write code to calculate this for ourselves without having to draw tables and do statistics by hand. We've already seen a bit of a brute-force method to do this, but it's not perfect by any means, so how can we do things better?

First, let's write a version of this formula which might be a bit easier to use in python. Since statistics can often be difficult to deal with in programming, I've come up with a formula to make this particular problem a lot easier for us.

We're gonna need to look at a special formula for dice rolls; it looks like the following:

$$E(x) = \sum_{n=1}^{diceNum} 0.5 + \frac{m}{2}$$

Where "m" is the maximum. Instead of finding the probability for each individual item in the list, we recognise that each item in the list has a pair that it adds up to to become ½ of the maximum number of dice, except for, of course, the number in the very middle, if a whole number exists there.

After that, we also recognise that we can sum 0.5 with itself a number of times equal to diceNum can provide us with the roll that is half of the maximum. It turns out, when we add these two values together, we get a formula that allows us to calculate the chance of getting any result from a dice roll.

We can actually derive this one purely off of thinking hard about it; 1d12 has 12 options, from 1-12, and the sum of numbers from 1-12 is 78. That divided by 12, in order to get the mean value, is 6.5. Likewise, 2d6 is made up of 2 individual d6. If we add up the values for each of those, we get a total of 42, which, when divided by 6, to get the average, we get 7. We can test this with 3d4 as well; 3d4 gives 3 separate d4 outputs, which, when summed together, give us a grand total of 30. That 30 is then divided by 4, making 7.5.

We can actually find the average even easier, by adding together the lowest and the highest possible roll, and then dividing by 2, and in fact, in a way, that's what our existing formula already does.

$$E(x) = \sum_{n=1}^{diceNum} 0.5 + \frac{m}{2} \Rightarrow \frac{diceNum}{2} + \frac{m}{2} \Rightarrow \frac{diceNum+m}{2}$$

Anyways, onto the code, where I use the summation-notation version of this formula to figure out the average of dice rolls.

```python
def getDice(dice):
    diceTotal = 0
    for i in range(len(dice)):
        diceTotal = diceTotal + dice[i]
    x=0.5*len(dice)+(diceTotal/2)
    return x

def getInputs():
    lst = []
    inp = input("Give me some numbers or 'q' to quit! >>")
    while inp != "q":
        if inp.isnumeric():
            lst = lst + [int(inp)]
        elif inp == "q":
            break
        else:
            pass
        inp = input("Give me some numbers or 'q' to quit! >>")

    return lst
while True:
    inputs = getInputs()
    print(getDice(inputs))
```

Here's where we're at so far: we use a function getInputs to get a set of inputs and a function getDice to do our calculations. getDice makes a variable called diceTotal, which is set to 0. This variable is then

added to through a loop, and then we do our formula.

This is great, but it's still missing something; so far, it reads

individual inputs until we get a "q". I want to change this; and in our

next chapter, we will discuss the basics of regex.

Chapter 17: Regex Strings

Let's be honest here, Regex can look intimidating with no prior context. I mean, just look at it:

```
re.compile('(\d+)d(\d+)(?:([+-])(\d+))?([ad])?')
```

But it's fine if you don't understand it; this chapter will give a very basic rundown of regex in python. Firstly, we need to understand regex patterns; and for this, I'm gonna use a really dumb example.

Let's say you're a super secret spy; You've got to deliver classified information to your contact, but there's always a chance that someone along the way will intercept it, so you've gotta be clever. You devise a plan; you won't actually tell or write it down, instead, you'll speak a series of codewords, which your spy will understand as prompts to check a certain place for certain things.

The analogy isn't great, but basically, regex allows us to parse strings for certain things. The above pattern actually looks for a dice roll in dice notation (IE 1d12+4d would be "1d12 + 4 at disadvantage").

Firstly, let's get a few regex bits out the way:

\d	Digit
\s	Whitespace
\n	New line
\w	Word characters (numbers and spaces)
\S	Not a whitespace
\b	Word boundary
a*, a+, a?	0 or more, 1 or more, 0 or 1
ab|cd	Match ab or cd
[abcd]	Any from a, b, c, or d
(abc)	Capture group
(?:abc)	Non-capturing group

Now, with just this, you can already figure out how my regex pattern works. However, for this, we only need '(\d+)d(\d+)(?:([+-])(\d+))?' This will allow us to check for a dice pattern in which any number of any sided dice is rolled, and any modifier is added to it.

Implementing it is actually really easy: After importing the regex module, all you really have to do is define your pattern and compile the pattern.

```python
import re
def getDice(diceStats):
    dice = int(diceStats.group(1))
    diceTotal = int(diceStats.group(2))*dice
    if (diceStats.group(4)==None):
        x=0.5*dice+(diceTotal/2)
    else:
        if(diceStats.group(3) == '-'):
            x=0.5*dice+(diceTotal/2)-int(diceStats.group(4))
        else:
            x=0.5*dice+(diceTotal/2)+int(diceStats.group(4))
    return x

def getInputs():
    dicePattern = re.compile('(\d+)d(\d+)(?:([+-])(\d+))?')
    inp = input("Give me a dice pattern")
    if re.search(dicePattern,inp):
        p1 = dicePattern.match(inp)
        return p1
    else:
        return None
while True:
    inputs = getInputs()
    if not (inputs == None):
        print(getDice(inputs))
```

As you can see, once we've defined our pattern, the module really does all of the work for us, so there's not really a ton of reason for me to explain it further, but… that's also the point of this book, so quickly; re.compile compiles a regex pattern, re.search searches for a regex pattern inside a text, Pattern.match matches a pattern in a string and, if preceded by an equals, makes a regex item with that pattern, and

Pattern.group gets a given capture group. We use all of these functions in tandem in order to process our input.

Now that we've gone over basic statistics, that leaves us with all of our topics covered;

Epilogue

Together, we have delved deep into the stories of numbers, taken a look at the intricacies of circuitry, learned the foundations of programming, come to understand how gravity pulls us together, and how to use probability to our advantage; but there's one more thing I'd like to leave you with, reader.

You've come far on your journey; you've learned to muse numbers, tell the tales of planets, and comprehend binary, but what I hope you have taken from the most is that there is always more to learn; don't stop with just my book, go find others to read, to enjoy as you have this one. Learn about what interests you, and go forth into the world slightly smarter and slightly wiser. Take care,

--Isaiah Watson,